TECHNOLOGY IN ACTION

MILITARY TECHNOLOGY

Chris Smith and Bernard Harbor

Wayland

Titles in this series

Aircraft Technology
Building Technology
Camera Technology
Car Technology
Energy Technology
Farming Technology
Food Technology
Information Technology
Medical Technology
Military Technology
Ship Technology
Spacecraft Technology
Sports Technology
Train Technology
TV and Video Technology
Undersea Technology

First published in 1991 by
Wayland (Publishers) Ltd
61 Western Road, Hove
East Sussex BN3 1JD, England

Edited by Roger Coote
Designed by David Armitage

Front cover The US B2 stealth bomber on a test flight in 1989. This plane uses technologies that make it almost invisible to radar. It represents the very latest in aircraft technology.

British Library Cataloguing in Publication Data
Smith, Chris
Military technology
1. Military equipment: weapons
I. Title. II. Harbor, Bernard III. Series.
623.4

ISBN 0-7502-0209-2

Typeset by Direct Image Photosetting Limited, Sussex, England
Printed in Italy by G. Canale & C.S.p.A., Turin
Bound in France by A.G.M.

Contents

1 What is military technology

We tend to think of military technology as a particular piece of equipment, such as a fighter aircraft, a warship, a missile or a tank, but most modern weapons are complex systems made up of a wide range of different technologies. Many of these technologies are extremely complicated; others are very familiar.

A military aircraft is, primarily, based on aerospace technology and is designed to drop bombs, fire missiles or fight other aircraft. To be able to do this, however, an aeroplane must incorporate other technologies, including sophisticated electronics such as display systems and advanced computers, as well as engineering technologies such as suspension, steering and braking systems.

Some technologies are used only in the military world; it is difficult to think of civil uses for a nuclear bomb or a machine-gun. However, there is a wide range of technologies that are used in both military and civil products. Computers, communication systems and many aerospace and shipbuilding technologies fall into this category and, of course, armed forces regularly use things that are familiar in our own lives — lorries, telephones, office equipment, and so on.

This book looks at some of the features of military technologies and what makes them different from civil technologies. To do this, we must first decide what *is* a military technology? The easiest way is to think of how things are

A US Navy pilot controls many separate and complex technologies, worth millions of dollars.

Hundreds of scientists, engineers and shipbuilders work to produce a military submarine.

used. We could say that anything designed for fighting wars or preventing them is a military technology. A tank, for example, incorporates a range of engineering and electronics technologies, some of which might have civilian uses, but we could say that it is a military technology because those technologies have been applied to produce a weapons system.

Another way of recognizing military technologies is to look at the way in which they differ from similar civil technologies. Military telephones do much the same job as those in our homes, but a military telephone system is made with the particular requirements of the armed forces in mind. For example, it might be mobile, or very tough. It will certainly include sophisticated codes to ensure that enemies cannot tap into the system and hear what is being said. These features are unnecessary for most civil telephones.

Military technologies are also developed and manufactured in different ways from civil technologies. The customer — that is the army, navy or air force — usually pays for the technology to be developed using money it receives from the government. The military tells industry what 'mission' or 'military requirement' it wants to be able to achieve and industry then tells the armed forces how it could be done. Usually, the armed forces then pay for industry to develop the necessary technologies; this is called 'research and development', or 'R. & D.'. Once the technology has been developed the armed forces will pay for the product to be manufactured. When this 'production' phase has been completed, the equipment will be delivered to the armed forces who can begin to train soldiers to use it; this is called 'deployment'.

Obviously, this is a very different process from the civil sector. It is rather like going to an electrical shop and paying for the most modern stereo system, and then waiting several years before delivery while it is designed and manufactured.

Above Military equipment often does the same job as civil systems, but has many more features.

Below Laser technology is a very important element of modern military electronics.

The armed forces rely on industry to make the weapons they need. In some countries, they also have their own laboratories where they research the most modern technologies. This means that military technologies are usually tailored to the exact specifications of the armed forces. Again, this ensures that military technologies are distinct from civil technologies.

Finally, military technology can be defined by who is allowed to buy and use it. Virtually any person or any country can have a civil product so long as they have enough money to pay for it, but there are strict controls on who can buy military technologies. In some countries, civilians are allowed to own guns and rifles, although they usually need a licence, but civilians are not allowed to own fighter aircraft, tanks or nuclear weapons. Similarly, those countries which make and own the most sophisticated weapons tend to restrict their sales to other countries. For example, an international treaty exists to restrict the number of countries that are allowed to own nuclear weapons. There are also some restrictions on international sales of conventional weapons and ballistic missiles.

Part of the barrel of a 'supergun' that was ordered by Iraq but seized by UK Customs. British arms exports to Iraq are illegal.

2 Mobility, firepower and protection

As we have seen, military technology is developed and produced in a very different way from civil technology. This is largely because defence systems are designed to be extremely competitive.

People have different reasons for buying a car — long-distance driving, shopping trips or to impress friends, perhaps — which will help them to decide what type of car to buy and how much money to spend. In the military world, the choices are very different. Money is less of a consideration, because the armed forces usually have huge resources at their disposal — far greater, for example, than health or education services. This is especially true in the Third World where welfare expenditure is often extremely low.

In order to decide what equipment they need, military planners will look in two directions. First they will assess what industry is capable of producing, both now and in the future. Second, they will look at what their enemies are buying and producing and attempt to match their capability. The result is known as an arms race.

Modern tanks are fast and manoeuvrable but they have become more and more vulnerable to anti-tank weapons.

A US F-15 Eagle. Modern fighter aircraft use afterburners to give them extra acceleration on take-off.

Then the planners will decide on the specific weapons systems they want, such as a tank, a ship or a plane. In making their decisions, the planners study each weapons system in terms of its mobility, protection and firepower.

Mobility gives a weapons system the speed and manoeuvrability to attack an enemy using as much surprise as possible or to escape from dangerous situations. If mobility is to be increased, more power is needed, and the power available depends on both the thrust of the engine and the weight of the body, or weapons platform. New metals and plastics, which are very light and strong, help to keep the weight down. Advanced engines are designed to give high levels of acceleration in relation to weight, which is known as the thrust to weight ratio. In the engines of fighter aircraft, for example, a device called an afterburner helps to produce intense acceleration at over 2,400 km/h, but only for a very short time because it uses a great deal of fuel.

This tomb of a Roman centurion shows the kind of armour worn in the first century AD.

Mobility does not depend just on speed, but also on manoeuvrability — the ability to turn in tight circles, avoid enemy fire and keep the enemy unsure about the exact position of a weapons system.

Protection can take several forms, depending upon the system. Ever since Roman times, protection in the form of armour has been very important, though sometimes quite basic. For centuries it depended upon materials that could withstand the impact and force of projectiles and explosives. In the modern era, the approach to protection has become wider. For example, anti-missile missiles are now used. Tank armour has become very sophisticated and its purpose is to deflect, absorb or destroy anti-tank missiles. Modern aircraft and ships tend to rely upon electronic systems to give them advance warning of an enemy approach.

Firepower is required to destroy opposing weapons systems or to cause harm to people or property. As nuclear weapons show, the effects of firepower are not always limited to opposing armed forces but can also affect civilians. The main elements of firepower are the speed and accuracy with which it is 'delivered' and the force of the explosion on impact. In response to advances in tank technology, anti-tank missiles have become more and more powerful. For example, a high-energy anti-tank missile delivers a jet of molten metal travelling at about 1,000 kilometres per hour and another anti-tank weapon, known as an 'armour-piercing discarding sabot', can penetrate up to a metre of armour. As firepower increases and becomes more able to find a tank's weak spots, such as the turret, tanks have become easier to destroy. Because of this, tanks will probably be used less in the future.

The ideas of mobility, firepower and protection are becoming increasingly linked together. However, it has always been the case that attention to one has meant less of another. A battleship has tremendous firepower but it also relies upon other ships and submarines for protection. Bomber aircraft have comparatively little mobility and very little protection and depend upon what is known as 'top cover' from fighter aircraft. Tanks are well protected but this makes them larger, heavier and slower.

Despite the sophistication of modern weapons systems, they can and quite often do go badly wrong. Some break down frequently because they are so complex; it only takes one of the thousands of interlocking parts to malfunction to turn the most advanced system into the most expensive target. Military planners are now paying greater attention to making weapons systems simpler and more durable.

Above Sophisticated weapons are often small enough to be carried on the back of a jeep.

Below This photograph of a Soviet military installation was taken from a space satellite orbiting the earth.

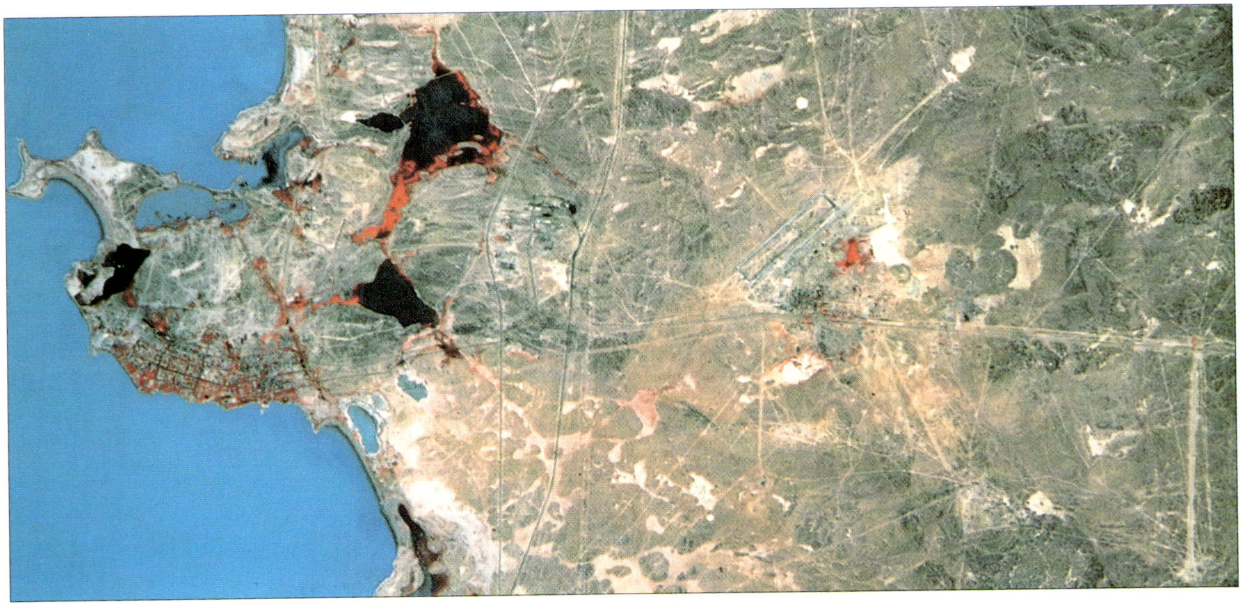

Military technologies are not only used for destructive purposes; some of them also have civil applications. Air, sea and land rescue services rely heavily on expertise and technologies developed originally for the armed forces, such as radar, flares and helicopters. In time, it will be possible to use satellite technology, originally developed for tracking the movements of far-off enemies, to give advance warning of droughts or floods.

Some technologies developed for the armed forces also have other uses – like this rescue helicopter.

3 Land warfare

Over the years warfare on land has changed greatly, although the physical size of national armies has changed little — they are still large. Throughout the world there are some 18 million full-time soldiers; both China and India have armies of over 1 million soldiers.

Land warfare has changed mainly as a result of the introduction of new technologies or innovations in existing ones. The first use of the longbow at the battle of Crécy in 1346, the use of repeating guns during the American civil war of 1861-65, and the discovery of the power of anti-tank missiles in the 1974 Arab-Israeli War are some of the important landmarks in the history of warfare.

In addition, land warfare has been changed by technological advances that did not begin as military projects. The invention of the railway was slow to excite military planners until they saw its potential for moving large numbers of

India has one of the largest land armies in the world with over a million soldiers.

men and equipment rapidly over long distances. Ever since the days when smoke signals were used, information technology has been of vital importance to military decision-makers. The telegraph first revolutionized communications, and the modern equivalent is the microchip, which permits information to be transmitted, sorted and analysed in fractions of a second. This is quite an improvement on the two-tonne, horse-drawn, radio transmitter/receiver used by the British Army in 1911.

During the First World War of 1914 to 1918, the use of the machine-gun and barbed wire gave the advantage to defending forces rather than attackers. The invention of the tank, an offensive weapon, broke the deadlock as a means of advancing and holding territory. After the War a great deal of attention was focused upon the tank and the role of armour in general. In recent years, tank technology has probably almost reached its limit, especially with regard to armour. Efforts are now directed at using tanks and artillery more effectively by exploiting information technology.

Great effort has also gone into developing anti-tank weapons, such as mines, missiles and guns. Missiles are especially important. In the case of the American TOW, which stands for 'tube-launched, optically sighted, wire-guided', the gunner sights his target through a telescope and fires. As the missile flies towards the target, control wires connecting the missile and the launcher unwind. In the tail of the missile are an infra-red source and an infra-red tracker, which make it possible for the gunner to send steering commands to the missile's fins and keep it on course. Another example is the American Maverick air-launched anti-tank missile which is steered by a television camera in its nose. Signals are sent back to the aircraft, and the pilot

The use of the machine-gun in the First World War completely changed the nature of land warfare.

Left A wire-guided anti-tank missile is launched. The operator can 'steer' the missile towards its target using control wires.

Below A Bofors mobile artillery weapon suffers a devastating direct hit by an anti-tank missile in a test firing.

then activates a device called a 'contrast seeker'. The missile compares the TV picture of its target with the image stored in its computer memory, and automatically adjusts its flight path. The pilot of the aircraft can then turn away without having strayed too close to enemy fire. These are known as 'smart' missiles because they do so much on their own and require little input from the operator.

Munitions have been improved in similar ways. Artillery fire can now be co-ordinated by computers to give greater accuracy. Bombs can be guided by fixing a laser beam on to the target. A metal known as depleted uranium is so hard that when used in a projectile it can penetrate even the thickest armour. It is also self-igniting and can have awful after-effects, particularly if it succeeds in penetrating tank armour and finds its way into the cramped space inside.

In the future, land warfare will be increasingly linked with air power. Missiles and artillery with great range and accuracy will be able to destroy tanks before they reach the 'forward edge' of the battlefield, where most of the fighting takes place. Alternatively, the same systems can be used to make craters in a runway so that enemy planes cannot take off, or to attack ammunition supply depots, repair workshops or key bridges.

Another trend in land warfare is towards the use of systems without pilots. In modern battles, Remotely Piloted Vehicles, or RPVs, will be used to seek out targets. Robots will be used for clearing mines and transporting ammunition and they will drive wheeled and tracked vehicles. As computers become more powerful, they will be used to help armed forces decide how to use their firepower and where to place troops and equipment.

4 Naval technology

US Navy ships escort an oil-tanker through the Persian Gulf. Keeping sea lanes open is an important job for modern navies.

The sea has naturally played a vital role in transport, trade and communications since ancient times because it covers almost three-quarters of the earth's surface. Its strategic importance has meant that the sea has often been both a cause of conflict and an arena for war. Without naval forces, invasion, overseas intervention and the supply and protection of overseas territories would be impossible.

Navies are used for many roles. They are used to protect important trade routes, such as the Arabian Gulf, where the USA and other Western countries have sometimes sent their navies to ensure that oil supplies get through. They are used to support military operations by carrying soldiers, weapons and other supplies across the world. They are also involved in naval warfare, fighting other ships and submarines. In recent years, navies have played an increasing role in preparing for nuclear war.

Naval warfare has been influenced by technology for thousands of years, and never more so than in the last 50 years. Since the Second World War, new technologies have changed naval warfare almost out of all recognition.

For centuries, the warship was the central element of naval warfare. Following the invention of the aeroplane, and its adaptation for use at sea, the role of the warship has gradually declined. The first aircraft-carrier — effectively a floating airfield — was invented in Japan during the late 1930s, and by the end of the Second World War it had become the most important part of any navy. Modern navies depend as much on their aircraft as on their surface ships because of their role in surveillance and attack.

However, in terms of importance, the aircraft-carrier was soon overtaken by the submarine. With the development of electrical storage batteries, petrol engines and nuclear reactors, it became possible to develop submarines which could stay underwater for long periods of time. Under water, submarines were invisible to surface ships, making them safe from attack and giving them a great advantage over surface ships in battle. This led to the development of a range of technologies aimed at detecting and destroying submarines.

For centuries, the warship was the central element of naval warfare. Today the submarine has taken over this role.

Right A US F-14 aircraft taking off from an aircraft-carrier.

Below The US Navy ship *Vincennes* firing anti-submarine weapons.

Modern sea vessels depend on powerful engines to move at great speed. Most submarines are propelled by diesel or electric motors, driven by batteries or, more recently, by nuclear power. A small nuclear generator provides almost unlimited power, enabling a submarine to travel at high speed without having to surface for months. It can also recirculate air and process sea-water for drinking.

Although they are extremely powerful, nuclear generators are also very large, heavy and expensive. So, only the very largest surface ships are nuclear powered; most use jet engines.

5 Sea weapons

Naval vessels are really only platforms for carrying weapons. Ships have carried guns and cannons for centuries. Although modern versions are highly sophisticated, these days the missile has become the most important sea weapon. Missiles are self-propelled guided weapons. They have their own motors to propel them and are 'steered' to their targets by complex guidance systems. Some guidance systems are based on radar; they send out electromagnetic signals which 'bounce' off the target and tell the missile where the target is. Other guidance systems detect heat and, in this way, missiles are attracted to their targets' exhaust systems.

Modern torpedoes are underwater missiles designed either to destroy submarines or to be fired from submarines at surface ships. They are guided in a similar way to missiles, except that sonar is used instead of radar.

A radar operator in the US Marines. Radar is an essential technology for navigation and for locating the enemy.

Above A torpedo fired from a submarine makes a deadly impact on a surface ship.

Below A French Exocet missile is fired from the deck of a warship. One direct hit from an Exocet can destroy a ship.

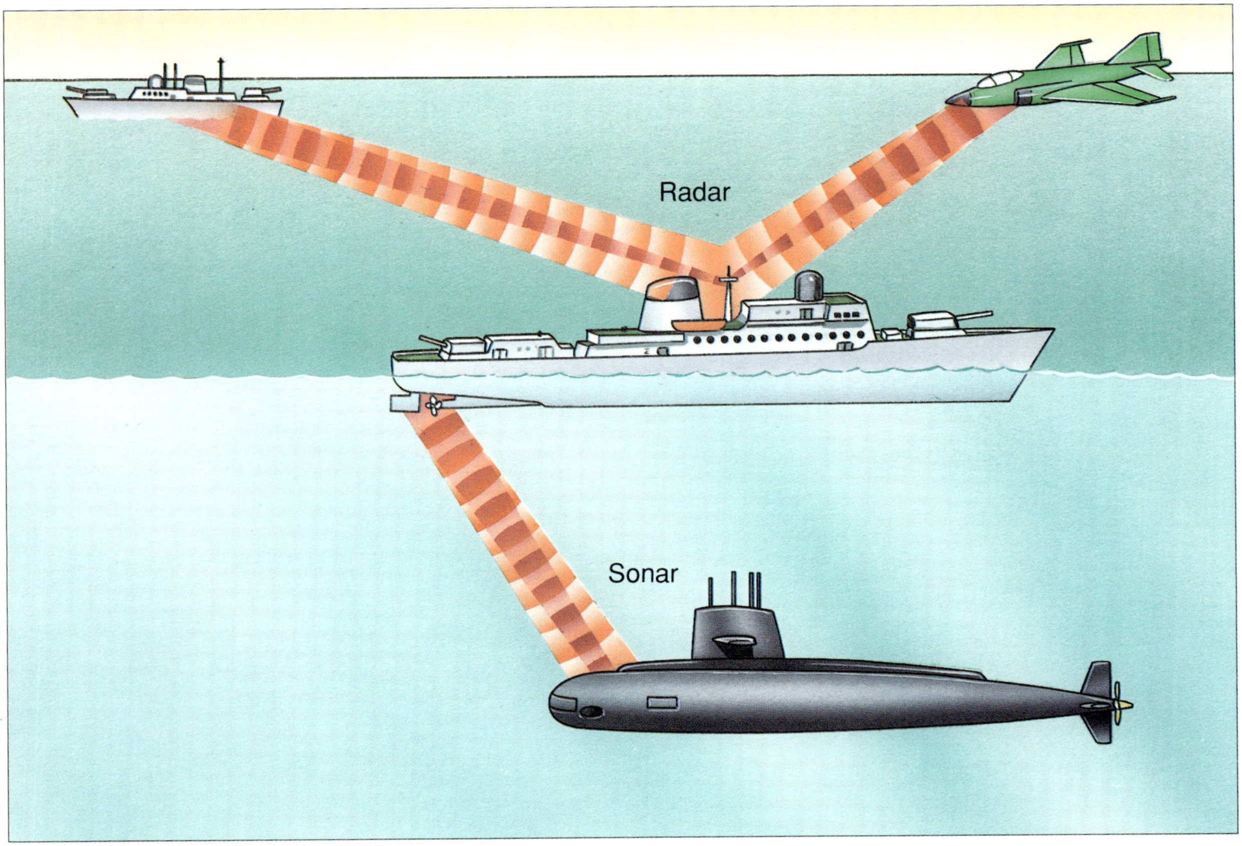

Radar uses radio waves to detect surface ships and aircraft, while sonar uses sound waves to locate objects underwater.

Because of their accuracy, modern missiles and torpedoes pose a major threat to surface ships and aircraft. Furthermore, although these weapons are expensive, they are much cheaper than their targets. The 'Exocet' missile which destroyed the British ship HMS *Sheffield* during the Falklands War in 1982, cost US$250,000 — about one-thousandth of the cost of a modern warship.

No matter how powerful a weapon is, it is of no use unless its target can be located accurately. Years ago, naval targets were detected simply by sight. These days, however, sophisticated technologies can identify targets hundreds of kilometres away.

Radar, which was invented in 1938, detects objects by sending out an electromagnetic signal which then bounces off an object back in the direction of the transmitter. By measuring the time it takes for the signal to return, the position of the target can be calculated.

Electromagnetic signals do not travel through water, and so radar is useless against submarines unless they are on the surface. However, the development of any military technology leads to the search for its antidote. Sonar, the underwater equivalent of radar, uses sound waves which travel well through water. The first 'passive' sonar systems simply listened for the noises given off by passing submarines. Today, 'active' sonar, which transmits an acoustic signal and waits for it to bounce back from its target, is used.

The main drawback to radar and 'active' sonar is that it is possible for enemies to detect the source of the signals. A ship or submarine using

Poseidon nuclear missiles are launched from a Polaris submarine. Nuclear submarines are very difficult for an enemy to detect.

radar or sonar can thus be detected and the hunter becomes the hunted.

Nuclear weapons are playing an increasing role in naval technology. Small nuclear warheads can be attached to almost any missile, but large 'strategic' nuclear missiles — the type which would be targeted on cities and military bases rather than ships or aircraft — are now carried on 'Ballistic Missile Submarines' known as SSBNs. These SSBNs can stay submerged and undetected for very long periods of time, which makes them much less vulnerable to attack than nuclear weapons based on land or on aircraft. For this reason, submarine-based nuclear weapons now play an important part in nuclear deterrence. Governments know that, no matter how much damage they inflict on another country and its defences, there will always be a submarine to retaliate with nuclear weapons. In theory, enemies could attempt to destroy nuclear submarines along with other nuclear systems, but this is very difficult in practice.

6 Military aerospace

At the outbreak of the First World War, military planners were still unsure of the usefulness of aircraft in warfare. Although hot-air balloons were often used for gathering information, the aeroplane was still new, and its potential unknown.

Modern land armies are useless without the 'cover' provided by military aircraft, and aerospace technology has provided a whole new range of weapons and military systems — helicopters, missiles, unpiloted aircraft and space systems.

Military aeroplanes were first used in the First World War.

The earliest use for military aircraft, and one which they still perform, was for gathering information, usually called 'reconnaissance'. Advances in engine technology and design made aircraft faster and more manoeuvrable, and enabled them to carry out other tasks. By the end of the First World War, they were already used in air-to-air combat or 'dog-fighting', and for dropping bombs.

The military and civil potential for using aircraft encouraged the rapid development of aerospace technology after the First World War. Early aeroplanes used propellers to drive them, and the speed of the plane was limited by the speed at which the blades of the propellers could turn. Much greater speeds were possible with the development of the gas turbine or jet engine. Instead of using propellers, a jet engine pushes an aircraft along by forcing air from the back of the engine at great speed. The first jet fighter, the Heinkel HE 176, was deployed by Germany in 1944, during the Second World War.

The Messerschmitt ME 262 was one of the first jet fighter aircraft to be developed.

A US Navy helicopter fires infra-red decoy flares. Their purpose is to draw heat-seeking missiles away from the helicopter.

The speed and manoeuvrability of military aircraft was further improved by the development of lighter materials and advances in aerodynamics, the science of flight. New aircraft shapes were designed to reduce the resistance or 'drag' of a plane as it flew through the air, enabling it to travel much faster. Flying at greater speeds put much more strain on an aircraft's structure, so new materials and designs had to be developed to increase its strength.

The 1950s saw the emergence of helicopters, which are sometimes called 'rotary winged' aircraft because they are lifted into the air by their rotating blades. As the blades turn, air flows around them and lifts the helicopter into the air. Helicopters are now commonly used for reconnaissance, transporting troops and firing missiles at tanks and other targets.

Two US B-52 bombers dropping large quantities of bombs over Vietnam during the Vietnam War.

Like other weapons systems, military aircraft were enhanced by the development of radar, which can be used both for navigation and to locate enemy targets. Computer technologies have also become increasingly important, especially in fighter aircraft. The most modern fighters are called 'fly-by-wire' systems because they are so complex it is impossible to fly them without the aid of computers. Modern pilots have to be expert computer operators.

Technological advances have greatly increased the military potential of aircraft. Modern fighters can travel and fight at speeds undreamt of by First World War pilots, and bombers can now carry huge loads over long distances. During the Second World War, what is known as 'strategic bombing' was developed. Strategic bombing means dropping huge amounts of bombs on targets like military bases or industrial areas, which are vital to a country's ability to fight. During the same period, air forces also began to attack vulnerable civilian targets, particularly cities. Hundreds of thousands of civilians were killed during bombing raids on cities in the Second World War. Over 130,000 people were killed in bombing raids on the German city of Dresden in 1945 and, later that year, the US bomber *Enola Gay* dropped the first nuclear bomb on the Japanese city of Hiroshima, killing 79,000 civilians.

Recent technological advances have made missiles more effective, and aircraft have become increasingly vulnerable. As a result, there has been a lot of discussion about the possibility of replacing piloted aircraft with Remotely Piloted Vehicles, or RPVs. They are already used for a variety of tasks, especially reconnaissance, and the time may come when they replace aeroplanes in bombing and even fighting roles.

Remotely piloted aircraft like this one will play an increasing role in future wars.

7 Missiles

Missiles were developed from similar aerospace technologies to aircraft. Before the invention of missiles, armed forces could use only cannon, which fired projectiles with no motors or guidance systems of their own. They depended entirely on the cannon itself for speed, power and direction.

Modern missiles are, at least to some extent, self-propelled and guided. Ballistic missiles are powered by chemical motors in the first part of their flight and, for the final part, follow their set trajectory and are brought to earth by gravity.

A cruise missile is powered by an engine for almost all of its flight and has 'wings' to help it maintain height. It is almost like an unpiloted aircraft with a warhead at its tip. The most advanced cruise missiles — those armed with nuclear warheads — have what are called 'terrain contour matching' guidance systems. These can recognize the area over which a missile is flying by comparing the contours of the ground with a map of the same area stored in the missile's own computer memory. Although never used, and now banned in its nuclear form, cruise was said to be unreliable — its computer became confused if there was snow on the ground and it had difficulty avoiding obstacles such as electricity pylons.

Missiles launched from aircraft can be used for fighting other aircraft, ships and tanks, as well as for attacking military or civilian targets on land. Ground-launched missiles can also be used against aircraft and, when coupled with sophisticated radar, can make very effective 'air defence systems'.

Scientists and engineers have tried to counteract air defence systems in two ways. The development of 'stand off' missiles, which can

Test firing a US Patriot anti-missile missile, which proved highly effective in the Gulf War of 1991.

This soldier's job is to aim, fire and control Rapier missiles. The aiming device is fitted into his special helmet.

be launched a long way from the target, has enabled some aircraft to deliver their weapons without flying across air defence systems. Also, with the aid of new materials and new aerodynamic designs, aircraft have been developed which are almost invisible to radar. These aircraft, the most modern in the world, are called 'stealth' aircraft. Only two types have so far been developed – a fighter and a bomber – both by the USA.

8 The militarization of space

Many of the advances in space technology have been paid for by military money because the potential for the military use of space is enormous. The military make extensive use of satellites. These devices are launched into space with rockets and are then placed in orbit around the earth, with their sensors pointed towards it.

Satellites can be used for a number of tasks. They can relay pictures to ground stations, for use in reconnaissance, navigation, targeting and early warning of attack. They can also be used for communications, as messages can be sent from one part of the world to another via satellite.

The growing use of satellites has led to the

A communications satellite is launched from the space shuttle *Discovery*. Many spaceflights have a military purpose.

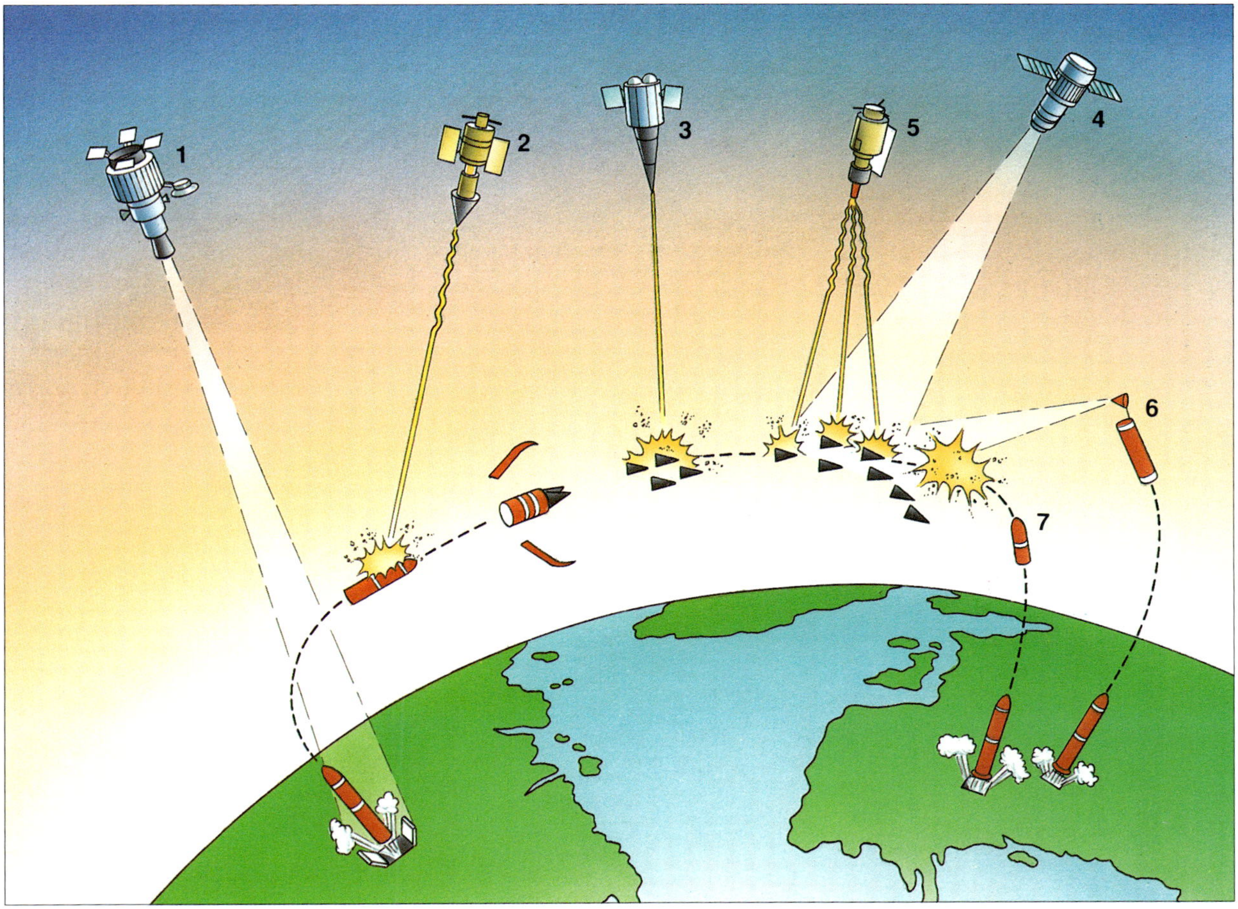

The 'Star Wars' defence system would locate and destroy enemy nuclear missiles using early-warning satellites (1), lasers (2 and 5), electro-magnetic weapons (3), tracking satellites (4), infra-red probes (6) and non-nuclear missiles (7).

development of anti-satellite weapons, or ASATs. The most simple is what is called a 'hunter killer' — a satellite which gets close to another and then explodes. Missiles launched from certain high altitude aircraft, or from rockets launched from the earth, can also be used as ASATs.

The most ambitious concept of war in space is the 'Star Wars' programme, devised by the USA. Development of the Strategic Defense Initiative, nicknamed 'Star Wars', began in 1983. The programme was designed to create a system of defence which would track, intercept and destroy incoming nuclear weapons and thereby give the USA protection from nuclear attack. By 1986 the USA was spending $3.7 billion per year on the programme. Research was carried out on powerful lasers, to be based both on the ground and in space. Scientists also investigated particle beam weapons, which send out a beam of particles to destroy a target by pulling apart the atoms of its structure. The programme has now been greatly reduced due to its high cost and questions over whether it would actually work. Attention is now being focused on the so-called 'Brilliant Pebbles' concept — a plan to place in orbit several thousand rockets capable of tracking and destroying incoming nuclear weapons.

9 Communications and computers

As we have seen, there is a vast range of sophisticated technologies and systems available to modern armies. To use these technologies on the battlefield, it is necessary to assess the strength and intentions of the opposition and to decide on a response. Because modern weapons are so sophisticated and destructive, it is essential to do this quickly and effectively.

A number of technologies have been developed to do this. Such technologies are not weapons — they do not have the capacity to kill or destroy — but they have become the cornerstone of modern warfare. Collectively, they are known as 'Command, Control, Communications and Intelligence' or 'C^3I'.

Military commanders must find out what forces the opposition has and where they are.

A mobile communications station in use with the US Army. Warfare is impossible without efficient communications.

Powerful computers help to make sense of all the information gathered on the battlefield and from space satellites.

The threat that they pose has to be calculated and decisions must be taken on how best to deal with them before any weapons can be used. Finally, the commanders must communicate their orders to the soldiers.

Finding out where opposition forces are is called 'intelligence'. This is done using sensors — such as radar and sonar — and reconnaissance — from aircraft, satellites and so on. Modern intelligence-gathering systems collect so much information that it would be impossible to make sense of it without the aid of powerful computer systems. These can process data to sort out which information is useful and can then analyse it. Finally, modern communications systems, which utilize telecommunications, radio, satellite and facsimile, relay information and orders to soldiers in the field.

An engineer tests a component of a sophisticated computer that uses Very High Speed Integrated Circuit (VHSIC) technology.

Effective C^3I is totally dependent upon advanced computer networks. The US Department of Defense has promoted the development of fast integrated circuits which can hold and assess vast amounts of data extremely quickly. They are called Very High Speed Integrated Circuits, or VHSIC. The USA has a vast world-wide C^3I network, called the US World-wide Military Command and Control System, which needs 90,000 people to operate and costs over US$1,000 million each year to run. The network uses 35 massive computers at 25 communications centres. It also includes 6 satellites, 600 facilities, and 100 ground terminals where information from the satellites is received, and it uses over 48 million kilometres of wire.

As with all other military technologies, advances in C^3I have led to the development of countermeasures. It is now possible to intercept messages, 'jam' systems to make them unworkable, or destroy them altogether. C^3I systems are networks made up of many parts, and each part is a potential target. A successful attack on a C^3I system can make military action impossible.

As computer technology advances, microchips — the parts of a computer which hold information and instructions — are getting much smaller. This means that more information can be held and processed by smaller computers. Together with advances in computer design and software, this is expected to lead to the development of 'supercomputers' with 1,000 times the power of today's most powerful computers. They may be capable of recognizing human voices, understanding language, and even enabling robots to 'see'.

The most ambitious computer development programmes hope to design computers that are able to think in the same way as human beings. Known as 'artificial intelligence', this idea is still science fiction but, if it could be developed, it would change the entire nature of warfare.

Modern computers can be made small and light enough to be carried around the battlefield.

10 Weapons of mass destruction

All weapons are intended to inflict damage and destruction upon people, property or other weapons. However, there are some weapons — nuclear, chemical and biological — that are designed to kill far more people than conventional weapons. They are known collectively as weapons of mass destruction.

There are two main types of nuclear weapons — the atom bomb and the hydrogen bomb. The first literally requires the splitting of an atom, usually of uranium or plutonium. The centre, or 'nucleus', of an atom is made up of tiny particles called protons and neutrons which are bound tightly together. If the bonds that link them can be broken by hitting the atom with another nucleus, a huge burst of energy is produced. Two or three more neutrons are detached when the atom is split and they in turn bombard and split other atoms. This leads to a chain reaction and an enormous explosion. This process is referred to as 'fission'.

The hydrogen bomb works in a different way, known as 'fusion'. Instead of splitting atoms of uranium or plutonium, which are very heavy, light hydrogen atoms are 'fused', or joined together. This does not happen naturally, and the only way to do it is by raising the temperature of hydrogen to 10 million degrees Celsius. At this temperature the hydrogen atoms are flying around so fast that they fuse. This leads to a chain reaction and a release of energy far greater even than in the fission process.

During the 1950s and 1960s, dozens of nuclear weapons tests were carried out on islands in the South Pacific Ocean.

Above The Japanese city of Hiroshima was devastated by a single nuclear bomb at the end of the Second World War.

Below These British soldiers in the First World War are wearing some of the first gas masks designed to protect them from poison gas.

Two atom bombs, one uranium and the other plutonium, were dropped on Japan during the Second World War, and the hydrogen bomb was invented shortly after. Nuclear weapons were first built by the two superpowers — the USA and the USSR — and then by Britain, France and China, to deter their opponents from attacking.

The most effective way of delivering nuclear weapons to their target is by missile. Ballistic missiles can travel great distances, even from one continent to another. They are fired into the atmosphere and then descend on to their target after the engine has cut out. Modern missiles can carry up to 12 nuclear warheads, which are fired when the missile has re-entered the atmosphere and can be directed at different targets. Smaller missiles with a shorter range are designed for use on the battlefield or within continents.

The other main types of weapons of mass destruction are chemical and biological weapons. Chemical weapons were first used

So-called 'Noddy' suits like these can protect soldiers from some of the effects of nuclear and chemical weapons.

during the First World War in the form of the lethal gases chlorine and phosgene and the less dangerous mustard gas. Modern chemical weapons use compounds known as nerve gases, which are invisible, have no smell and can kill in less than a minute. The gas prevents the movement of muscles in a person's body including the muscles that govern breathing, and this causes death by asphyxiation. Chemical weapons can be dropped from aircraft or delivered by artillery shells or missiles.

A major problem with chemical weapons has been the storage and transportation of such dangerous gases. To overcome this problem, the USA invented 'binary' weapons, which contain separate chemicals that are harmless on their own but lethal when mixed together. As the chemical weapon is fired, membranes separating the chemicals are broken, and the chemicals are mixed during flight to form a nerve gas.

Biological or germ weapons, which use bacteria and viruses to kill people, have been outlawed since 1972, although research still continues in the USA and the USSR. They were banned partly because they are such horrible weapons, and partly because they are difficult to control and can affect those who use them.

11 The effects of military technology

As military technology has developed, it has become much more destructive. The most frightening examples of modern military technology are weapons of mass destruction.

The use of a large number of nuclear weapons would bring an end to civilization as we know it. For many years it was thought that this meant that the opposing forces in the East and West would attack each other and blow up the world. Billions of people would die initially from the effects of blast, heat and radiation, and then from diseases such as cancer caused by fall-out. More recently, however, scientists have argued that even a more limited nuclear war could bring severe and permanent damage to the earth. It is thought that a nuclear war might produce a mass of dense cloud and dust, large enough to encircle the northern hemisphere and cut out almost all of the sun's energy for a week or more, causing temperatures to fall by between 15° and 30°C. People would die from the cold, as would the crops, trees, plants, animals and fish on which we depend. Eventually the southern hemisphere would be affected too, particularly if the clouds moved southwards.

Temperatures would start to rise slowly as the smoke thinned out, but it would be impossible for the world to return to normal again. Apart from the destruction of the planet's ecosystem, the ozone layer surrounding the earth would also have been destroyed, letting in lethal doses of the sun's ultraviolet radiation, which causes skin cancer.

Chemical and biological weapons would not cause damage on the same scale, but whole towns and regions could be destroyed if they were used in sufficient quantities.

It is often thought that there is a big difference between the power of weapons of mass destruction and that of conventional weapons, but this is not always true. Some conventional weapons can kill people over a wider area than small, 'tactical' nuclear weapons. For example, a fighter-bomber armed with a full load of about twenty high-explosive bombs is more than twice as lethal as a Lance nuclear missile.

US Navy planes dropping high-explosive bombs. Such bombs can be more devastating than nuclear weapons.

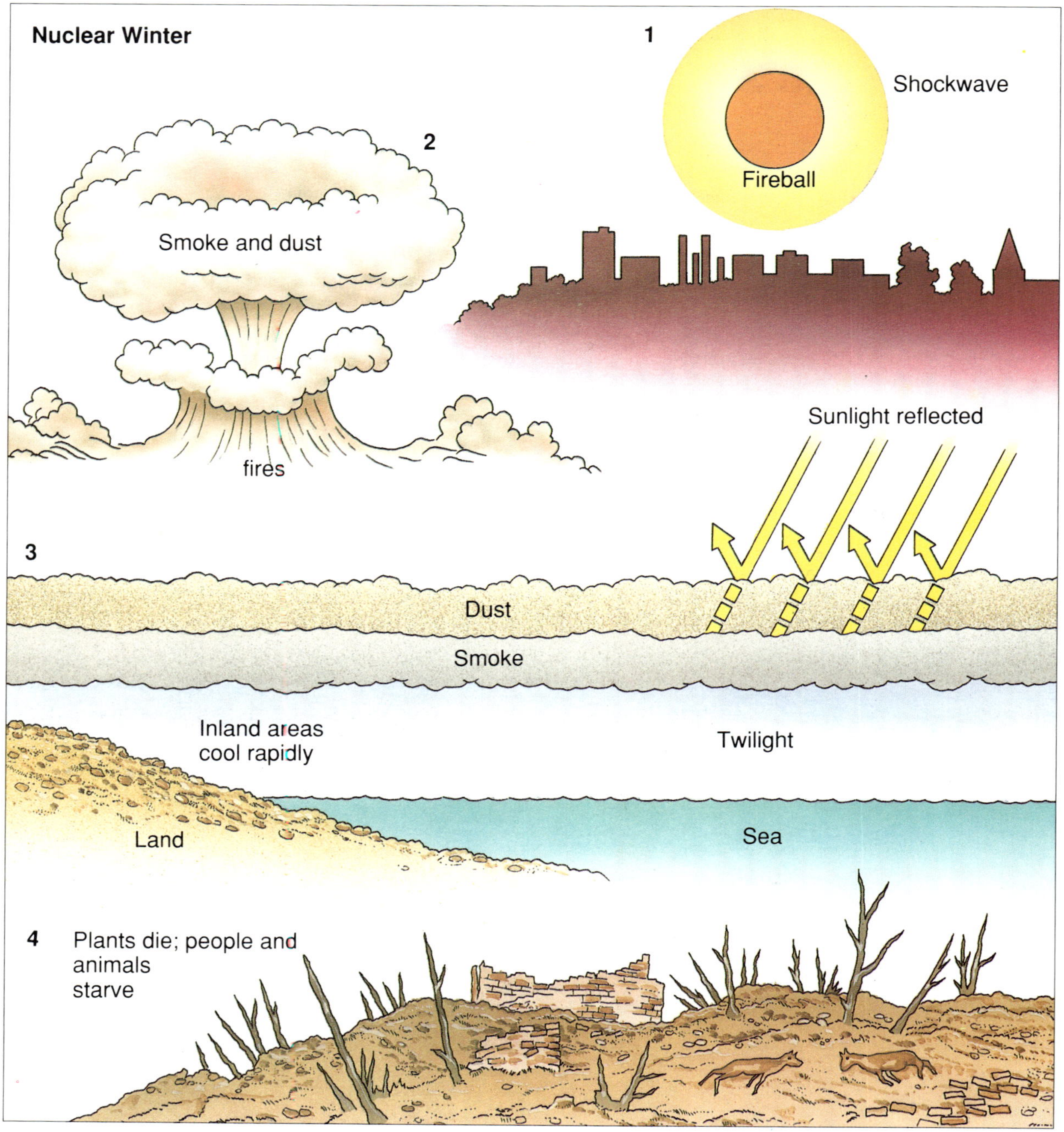

As weapons systems have become more sophisticated, they have become more manoeuvrable, better protected and more difficult to destroy. There are advanced missiles that can track and hit a moving target, but, more importantly, it is now possible to attack the entire area in which a system is manoeuvring. This can be done using fuel-air explosives, each of which consists of a batch of small bombs placed inside a single container. A number of containers are dropped from a slow-flying aircraft at about 500 metres above the ground. The containers open up at a pre-determined height, releasing the bombs on parachutes.

The bombs then float down in different directions. When they land, they explode and release ethylene oxide vapour which forms a cloud about 15 metres in diameter and 2.5 metres deep. The vapour is then ignited, causing a blast five times more powerful than the explosive TNT.

Other weapons are designed to injure people, and not necessarily to kill them. They are called 'anti-personnel weapons'. Bullets have been developed that explode inside a person's body and cause great damage to bones, internal organs and muscles, ensuring that, even if the victim is not killed, he or she will not recover sufficiently to fight again. A jelly-like substance called napalm is dropped in canisters from aircraft. When the canisters break open, the napalm ignites, forming a sheet of flame that causes terrible burns. High-explosive blast weapons kill people by causing a massive shock wave. This makes a victim's lungs collapse, forcing the escaping air up into the brain, causing brain damage or death. Other internal organs are also severely damaged.

Not all weapons are intended to harm people or destroy equipment. Some are designed to alter the area in which a conflict is taking place, often with disastrous effects on the environment. During the war in Vietnam, the Vietcong who had invaded the country were able to use the dense jungle as a protective cover. To counter this, the USA decided to destroy vast areas of the jungle. Between 1962 and 1971, almost 80 million litres of chemicals called defoliants and herbicides were used in Vietnam to remove ground cover and ruin the enemy's food supplies. One of the defoliants, Agent Orange, was particularly harmful and has since been identified as a cause of cancer, birth defects and a high incidence of still-born babies in Vietnam and among those who handled the chemical.

Also during the Vietnam War, attempts were made to cause rain to fall over certain areas of the country by dropping silver iodide and lead iodide from aircraft, in order to destroy food crops and make travel through the jungle extremely difficult. It may also be possible to cause tidal waves, snowstorms, earthquakes, volcanoes, avalanches and landslides for military purposes, but this type of activity has been outlawed since 1977.

Left The effects of defoliants used by US forces in the Vietnam War.

Below US planes spraying defoliation chemicals over Vietnamese forests in 1967.

12 What can go wrong?

Because military technologies are highly advanced and very expensive, it is tempting to assume that they work perfectly. In fact, the history of modern warfare has shown that military technologies are rather unreliable.

There are a number of reasons for this. Modern weapons systems are made up of thousands of parts, any of which might break down. If just one among thousands of working parts should fail, the whole system could stop working. It has been estimated that a piece of equipment fails on the US F-111D nuclear bomber every twelve minutes; each time the aircraft flies, it needs 98 person-hours of maintenance.

Malfunctions of weapons systems could clearly have serious consequences. More worrying still is the potential for C^3I systems to fail — particularly where nuclear weapons are concerned. In June 1980, failures in minor parts of American computer systems caused false alarms that a nuclear attack had started. Before the fault was located, bombers carrying nuclear weapons were prepared for take-off. There have been many other similar examples. As technological advances reduce the warning time before an attack, there is a greater risk that such malfunctions could start wars.

The most modern armies are rarely involved in wars, and so they have few opportunities to

This fighter aircraft has suffered considerable damage because of a simple malfunction in its landing gear.

Above In the 1980s the huge Soviet Army was unable to defeat the Afghan guerillas such as these.

Below Simple defects were discovered in Sea Harrier aircraft during the Falklands War in 1982.

test their weapons in the situations for which they are designed. This has often led to design faults in weapons systems going unnoticed, sometimes only being discovered in combat. When British Sea Harrier aircraft were deployed in the Falklands War, it was found that their windows steamed up making vision impossible. Such problems can sometimes be solved easily, but they could make weapons systems unusable in the short term.

Modern, high-technology armies have also found that they can be vulnerable to less well-equipped armies. Small, relatively cheap missiles can destroy expensive warships and aeroplanes. In 1967, Egyptian Styx missiles fired from a small patrol boat sank the Israeli destroyer *Eilat*. For all its technological superiority, the USA was unable to win the war in Vietnam against a peasant army, and the same was true of the USSR when it invaded Afghanistan.

Glossary

Acoustic To do with sound.
Aerodynamics The science of moving solid things, especially aircraft, through the air.
Aerospace To do with aeroplanes and flight.
Afterburner A mechanism designed to give jet engines a short but powerful burst of speed.
Aircraft-carrier A large ship from which aircraft can take off and land.
Antidote A means of counteracting or curing something.
Anti-missile missile A missile designed to destroy incoming enemy missiles.
Application The use that a thing can be put to.
Armed forces Military forces — including the army, navy and air force.
Armour-piercing discarding sabot An anti-tank weapon in which the warhead 'unwraps' as it flies to expose a hard inner bullet which can penetrate tank armour up to a metre thick.
Artillery A word for cannon or other mounted large guns.
Asphyxiation Suffocation; a lack of oxygen in the blood.
Ballistic A type of missile which is powered during the early part of its flight but is then drawn to earth by gravity.
Civil Not military.
Computer An electronic machine that makes calculations, and stores and analyses information.
Conventional A word to describe weapons that are not chemical, biological or nuclear.
Counteract Neutralize or hinder by taking opposite action.
Defoliant A chemical used to strip a plant of its leaves.
Depleted uranium A very hard metal which is a by-product of the process of producing nuclear warheads.
Deployment Putting a weapon into position ready for use.
Display system A system which shows information on a screen.
Dogfighting Air-to-air fighting between aircraft.
Drag The resistance felt by something moving through the air.
Ecosystem The way in which all the living and non-living things in an environment interact with each other.
Electromagnetic Magnetic forces produced by electricity.
Engineering The science concerned with the control, power and design of machines.
Facsimile A system by which copies of documents and pictures can be sent via telephone networks.
Fall-out The radioactive materials that fall to the ground after a nuclear explosion.
Firepower The destructive capability of a weapons system.
Generator A machine that produces gas, steam, electricity or other power.
Guidance system A means of guiding a missile to its target.
Herbicide A chemical used to kill plants.
Integrated circuit A small component of a computer which stores information.
Intelligence Information about an enemy's forces and intentions.
Jam To interfere with electronic equipment to stop it working.
Manoeuvrable Able to change direction quickly.
Microchip Another name for an integrated circuit.
Militarization Use for military means.
Mobility The ability to move.
Munitions A word for explosives contained in the warheads of a weapons system.

Navigation Finding direction.
Propellers The blades which propel an aircraft.
Propulsion Any means of moving something.
Reconnaissance Examining territory to discover the whereabouts and strength of enemy forces.
Retaliate To attack an enemy who has attacked you.
Role The job which a person, machine or other object is expected to do.
RPV A remotely piloted vehicle; an aircraft that has no pilot aboard but is directed from the ground.
Self-igniting The ability to burst into flames spontaneously.
Software The name for computer programs that tell the computer what to do.
Sophisticated Highly developed or complicated.
Specifications The description of features which a weapon or piece of equipment should include.
Star Wars The nickname given to the USA's project to design a protective shield against nuclear attack. The correct name is Strategic Defence Initiative.
Strategic Of vital importance to a country's security. Strategy is a word to describe the overall plan according to which a country might fight a war.
Submarine A naval vessel that can travel underwater.
Surveillance Close observation of an enemy or suspect.
Tactical A word describing weapons designed to carry out a specific task on the battlefield.
Telecommunications The technology of communicating over long distances.
TNT A type of high explosive.
Tolerance The ability, especially of aircraft, to withstand pressure.
Trajectory The path of an object, such as a missile, through the air.
Treaty An agreement between countries, sometimes to limit the number of weapons each country possesses.
VHSIC An integrated circuit that processes information at very high speeds.
Warhead The part of a missile that contains the explosive power.
Weapons system A platform, such as a vehicle, ship or aircraft, with military equipment in the form of guns, missiles, radar, and so on.

Further reading

Barnaby, F *The Automated Battlefield: New Technology in Modern Warfare* (Oxford University Press, 1987)
Brett, B *The Fighting Ship* (Oxford University Press, 1985)
Embleton, G A *Warfare in History* (Wayland, 1984)
Gibson, M *Twentieth Century Warfare* (Wayland, 1989)
Messenger, C *Tanks and Other Armoured Vehicles* (Warne, 1984)
Sheehan, M and Wylie, J *The Economist Pocket Guide to Defence* (Basil Blackwell, 1986)
Smith, C and Harbor, B *Nuclear Weapons* (Wayland, 1989)

Index

Picture acknowledgements

The publishers would like to thank the following for allowing their photographs to be used in this book: Ann and Bury Peerless 13; The Research House *front cover*, 4, 5, 6 (both), 8, 11, (top), 14, 15 (bottom), 16, 18 (both), 20, (both), 24, 25, 26, 27, 28, 32, 33, 34, 35, 37 (both), 39, 41 (right), 42, 43 (top): Rex Features 7, 9, 11 (bottom), 12, 15 (top), 19, 29, 30, 38, 43 (bottom); Topham 22, 23, 36, 41 (left); Wayland Picture Library 10, 17.
Artwork by Peter Bull.